DISCOVER!

STRANGE FISH

SEAHORSES ARE STRANGE

NATALIE HUMPHREY

Enslow PUBLISHING

I0880118

Please visit our website, www.enslow.com. For a free color catalog of all our high-quality books, call toll free 1-800-398-2504 or fax 1-877-980-4454.

Cataloging-in-Publication Data
Names: Humphrey, Natalie.
Title: Seahorses are strange / Natalie Humphrey.
Description: New York : Enslow Publishing, 2023. | Series: Strange fish | Includes glossary and index.
Identifiers: ISBN 9781978528468 (pbk.) | ISBN 9781978528482 (library bound) | ISBN 9781978528475 (6pack) | ISBN 9781978528499 (ebook)
Subjects: LCSH: Sea horses–Juvenile literature.
Classification: LCC QL638.S9 H86 2023 | DDC 597'.6798–dc23

Portions of this work were originally authored by Melissa Raé Shofner and published as *Sea Horses*.
All new materials this edition authored by Natalie Humphrey.

Published in 2023 by
Enslow Publishing
29 E. 21st Street
New York, NY 10010

Copyright © 2023 Enslow Publishing

Designer: Tanya Dellaccio
Editor: Natalie Humphrey

Photo credits: Cover Rich Carey/Shutterstock.com; p. 4 feathercollector/Shutterstock.com; p. 5 tank200bar/Shutterstock.com; p.7 Daniel Lamborn/Shutterstock.com; p. 9 Drew McArthur/Shutterstock.com; p. 11 P.Ubonwan/Shutterstock.com; p. 12 Zaher Eldin Shalata/Shutterstock.com; p. 13 GOLFX/Shutterstock.com; p. 15 Enessa Varnaeva/Shutterstock.com; p. 17 Bernard S Tjandra/Shutterstock.com; p. 19 Arunee Rodloy/Shutterstock.com; p. 21 Richard Whitcombe/Shutterstock.com.

All rights reserved. No part of this book may be reproduced in any form without permission in writing from the publisher, except by a reviewer.

Printed in the United States of America

CPSIA compliance information: Batch #CSENS23: For further information contact Enslow Publishing, New York, New York, at 1-800-398-2504.

Find us on

CONTENTS

Boldface words appear in Words to Know.

WATER HORSES

With a horse-like head and a mouth that never closes, seahorses are strange fish! Closely related to pipefish and sea dragons, seahorses are masters of **disguise**. Blending right into **coral reefs** and sea grasses, seahorses can disappear in plain sight!

SOME SEAHORSES CAN CHANGE COLOR TO HELP THEM BLEND IN.

AROUND THE WORLD

Most seahorses live in warm salty water around **coastlines**. They prefer water less than 65 feet (20 m) deep. A few species, or kinds, live near the United Kingdom in colder water. They can't live in just fresh water, but some kinds live in water that's a mix of fresh water and salt water.

SEAHORSES CAN MOVE THEIR EYES IN TWO DIFFERENT DIRECTIONS AT ONCE!

HOLDING ON

Seahorses have strange bodies. Instead of ribs like other fish, they have bony rings that make up their body and tail! Their tail is prehensile, which means it's able to hold on to things. They'll often wrap their tail around something to keep from drifting away!

SEAHORSES ARE USUALLY BETWEEN 0.6 TO 14 INCHES (1.5 TO 35.5 CM) LONG!

SLOW MOVING

Seahorses have small fins on their bodies to help them get around. The fin on their back, called the dorsal fin, helps them stay upright. The dorsal fin moves quickly back and forth, pushing them forward. The fins on the sides of their head help the seahorse turn. These are pectoral fins.

SEAHORSE BODY PARTS

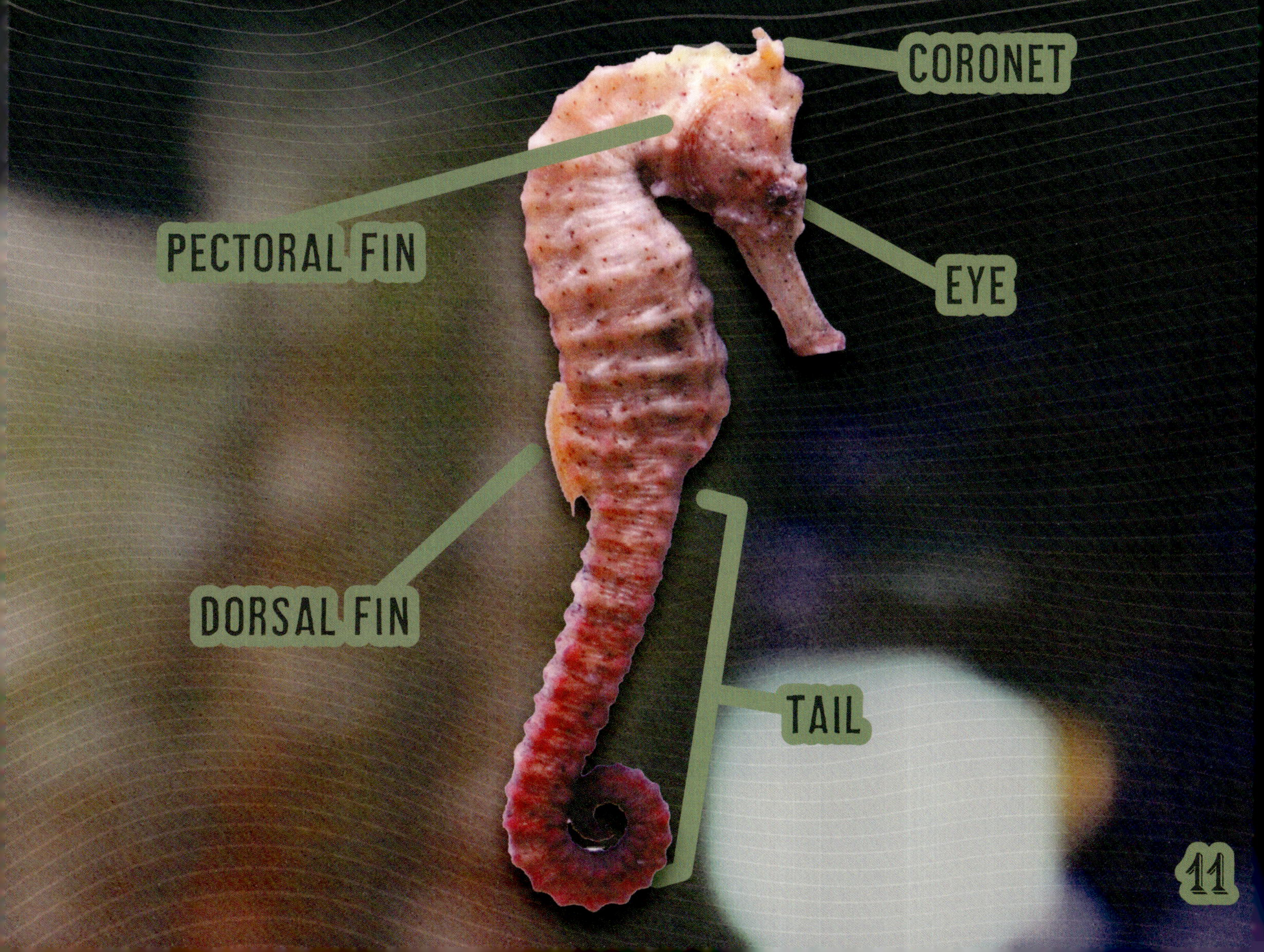

BONY BODIES

Seahorses might look the same, but scientists figured out a strange way to tell them apart! Certain species of seahorses have different numbers of bony body rings. Others have a bony bump on the top of their head called a coronet. No two coronets are the same!

THERE ARE 47 DIFFERENT KINDS OF SEAHORSES.

SO MANY SHRIMP

Seahorses don't have stomachs. Instead, they must eat all the time to have enough energy to stay alive. When they find a meal, they use their long, skinny **snout** to suck it out of the water like a straw! Seahorses eat small shrimp, tiny fish, and **plankton**.

ONE SEAHORSE CAN EAT 3,000 SHRIMP IN ONE DAY!

A BIG FAMILY

When seahorses are ready to **mate**, they are even stranger! The female seahorse places her eggs into a special pocket on the front of the male's body. The eggs stay inside their father's pouch for several weeks and then the babies hatch, or come out of the egg.

MOST SEAHORSES MATE FOR LIFE.

FISH FOOD

Because seahorses are so slow, they make the perfect snack for many hungry **predators**. Crabs, sharks, rays, birds, and even other fish like to eat seahorses. The only way seahorses can avoid being eaten is to hide. Many predators will swim right by a hidden seahorse!

Seahorses use **camouflage** to blend right into coral reefs.

PROTECT THE SEAHORSE!

One of the biggest threats to seahorses is people. Because they're so strange, many people try to keep wild seahorses as pets. Without special care, seahorses can't survive out of the wild. Coral reeds are also harmed by **pollution** and **climate change**. Without coral reefs, seahorses will have nowhere to live!

THE BEST WAY TO SEE SEAHORSES IS AT AN AQUARIUM.

WORDS TO KNOW

camouflage: Colors or shapes in animals that allow them to blend in with their surroundings.

climate change: Long-term change in Earth's climate, caused mainly by human activities such as burning oil and natural gas.

coastline: The land along the edge of a coast.

coral reef: An underwater hill made up of the hard parts of tiny sea animals.

disguise: The act of changing how one looks in order to hide.

mate: To come together to make babies.

plankton: A tiny plant or animal that floats in the ocean.

pollution: The action or process of making land, water, or air dirty and not safe to use.

predator: An animal that hunts other animals for food.

snout: An animal's nose and mouth that stick out.

FOR MORE INFORMATION

BOOKS

Duhaime, Darla. *Sea Horses*. Vero Beach, FL: Rourke Education Media, 2018.

Bodden, Valerie. *Seahorses*. Mankato, MN: Creative Education and Creative Paperbacks, 2023.

WEBSITES

A-Z Animals
a-z-animals.com/animals/seahorse/
Learn more interesting seahorse facts on this website.

National Geographic Kids
kids.nationalgeographic.com/animals/fish/facts/seahorse
Watch interesting videos and check out more fun facts about seahorses.

Publisher's note to educators and parents: Our editors have carefully reviewed these websites to ensure that they are suitable for students. Many websites change frequently, however, and we cannot guarantee that a site's future contents will continue to meet our high standards of quality and educational value. Be advised that students should be closely supervised whenever they access the internet.

INDEX